NATURE CLUB

SEASHORES

JOYCE POPE

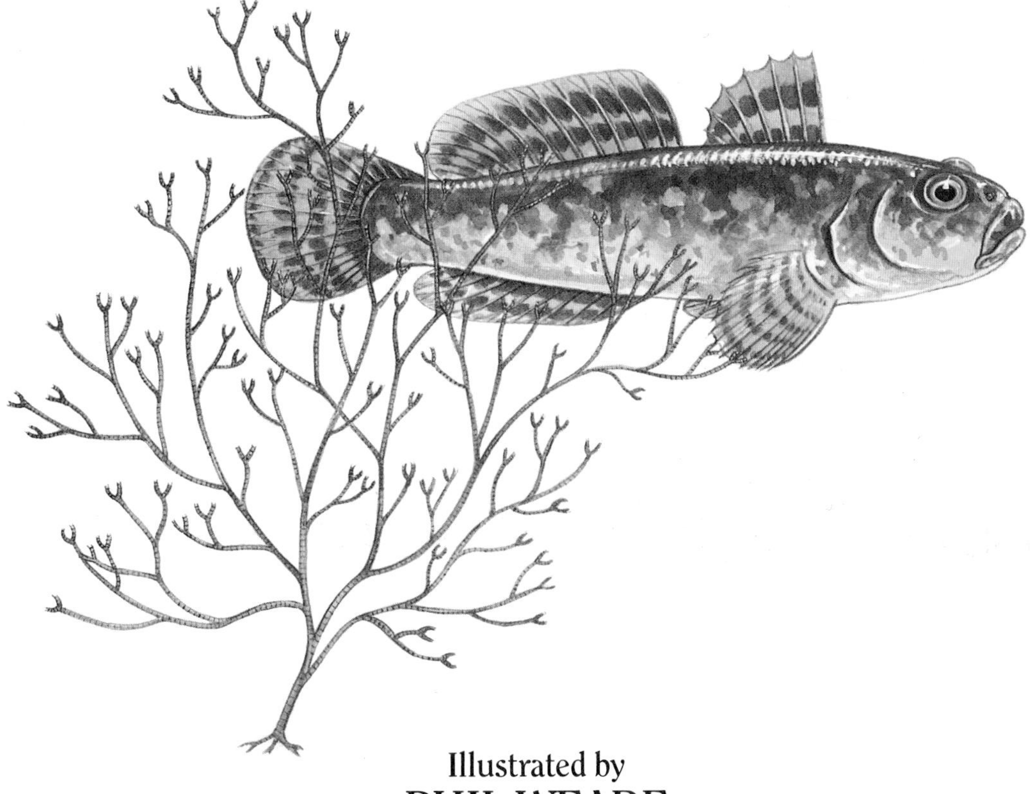

Illustrated by
PHIL WEARE

Troll Associates

Nature Club Notes

Though you may not know it, you are a member of a special club called the Nature Club. Anybody may join it. To be a member you just have to be interested in living things and want to know more about them.

Members of the Nature Club respect all living things. They look at and observe plants and animals but do not collect or kill them. If you take a magnifying glass or a bug box with you when you go out, you will be able to see the details of even quite small plants or shells, or tiny animals. There are many books that can help you to name what you have found and tell you something about it. Some are small enough to take with you, but many natural history books are big and heavy. Because of this, you should always take a notebook and pencil (or a pen that will not run if the paper gets wet) so that you can make a drawing of anything that you don't know. Don't say "But I can't draw" – even a very simple sketch can help you to identify your discovery later on.

As well as your notebook, your bag should contain a waterproof jacket and something to eat. It is silly to get cold or wet or hungry when you go out. Grownups are living things, too, so you should not worry them. Always tell your parents or a responsible adult where you are going and what time you are coming back.

The seashore has hidden dangers. Be careful at the sea's edge, or near deeper pools. Remember if you explore a beach at low tide that the sea can return unexpectedly. Even when you keep safety in mind there are always plenty of interesting places on a beach and lots of new creatures to be discovered.

Library of Congress Cataloging-in-Publication Data

Pope, Joyce.
Seashores / by Joyce Pope; illustrated by Phil Weare.
p. cm.—(The Nature club)
Summary: Describes the characteristics of various animals that live in or near the ocean.
ISBN 0-8167-1965-9 (lib. bdg.) ISBN 0-8167-1966-7 (pbk.)
1. Seashore ecology—Juvenile literature. [1. Marine animals.]
I. Weare, Phil, ill. II. Title. III. Series.
QH541.5.S35P63 1990
574.5'2638—dc20 89-20318

Published by Troll Associates, Mahwah, New Jersey 07430

Designed by Cooper Wilson, London
Design consultant James Marks

Printed in the U.S.A.

10 9 8 7 6 5 4 3 2 1

Contents

The Seashore

The seashore is where the ocean meets the land. It might be steep and rocky. Or it can be flat and sandy, pebbly, or muddy. But twice a day, at high tide, it is covered by the sea. At low tide, the sea drains away to show the sloping shore. Plants and animals that live on the beach have to be able to stand a life of both wetness and dryness. They are different from an animal of the open sea, which spends its whole life in the water, or land creatures that can only survive in the air.

Because the shore slopes, the plants and animals that live there are covered for varying lengths of time as the tide creeps up the beach. Some spend almost all of every day in the air, while others are dry for only a few minutes each month. As a result, you can see that the seashore is divided into zones, with different kinds of plants and animals in each as you go toward the open water.

A transect is a line map of the beach. To make one, simply fasten a length of string to a rock. Stretch the string toward the sea and secure the end with another rock. Record everything the string touches, making 1 inch in your notebook equal 2 feet on the string.

The tides vary from day to day during the month. When there is a new moon or a full moon, *spring tides* (the name has nothing to do with the season) occur. At these times, the tide comes in very high, then goes out very far. When the tide goes out, you have the best chance to see creatures that are usually covered by water.

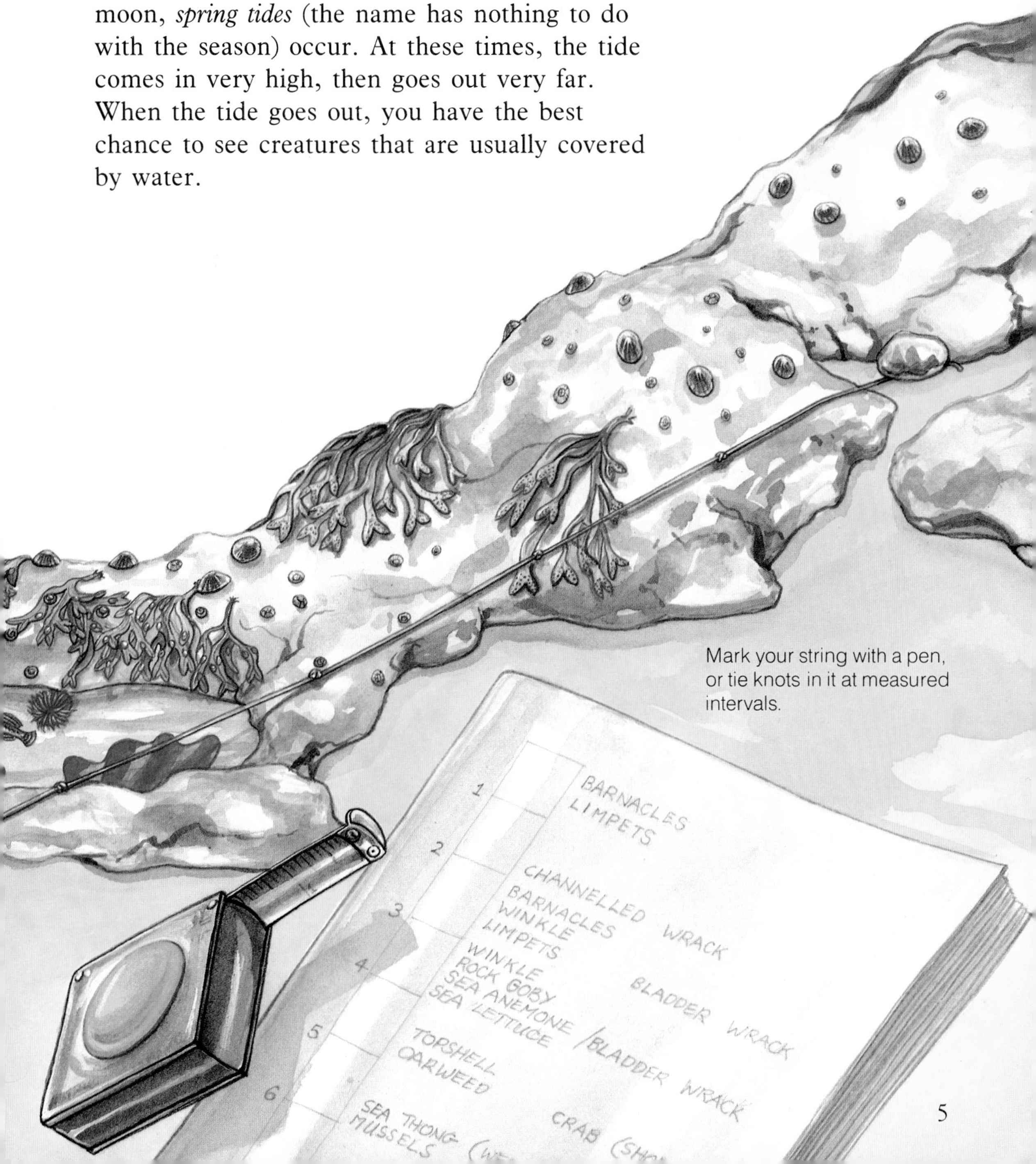

Mark your string with a pen, or tie knots in it at measured intervals.

Where Do Seashore Animals Live?

When you first go to the seashore, you don't usually see many animals, since most of them are small. Besides, most of them are hidden. This is because almost all of the creatures of the beach breathe through gills. They take in the oxygen brought to them by the water at high tide. But when you visit them at low tide they are resting and can manage with very little oxygen – perhaps some trapped in a tiny drop of water in their shells. Also, many of them make themselves safe from enemies by hiding in the sand or keeping out of sight among the seaweed.

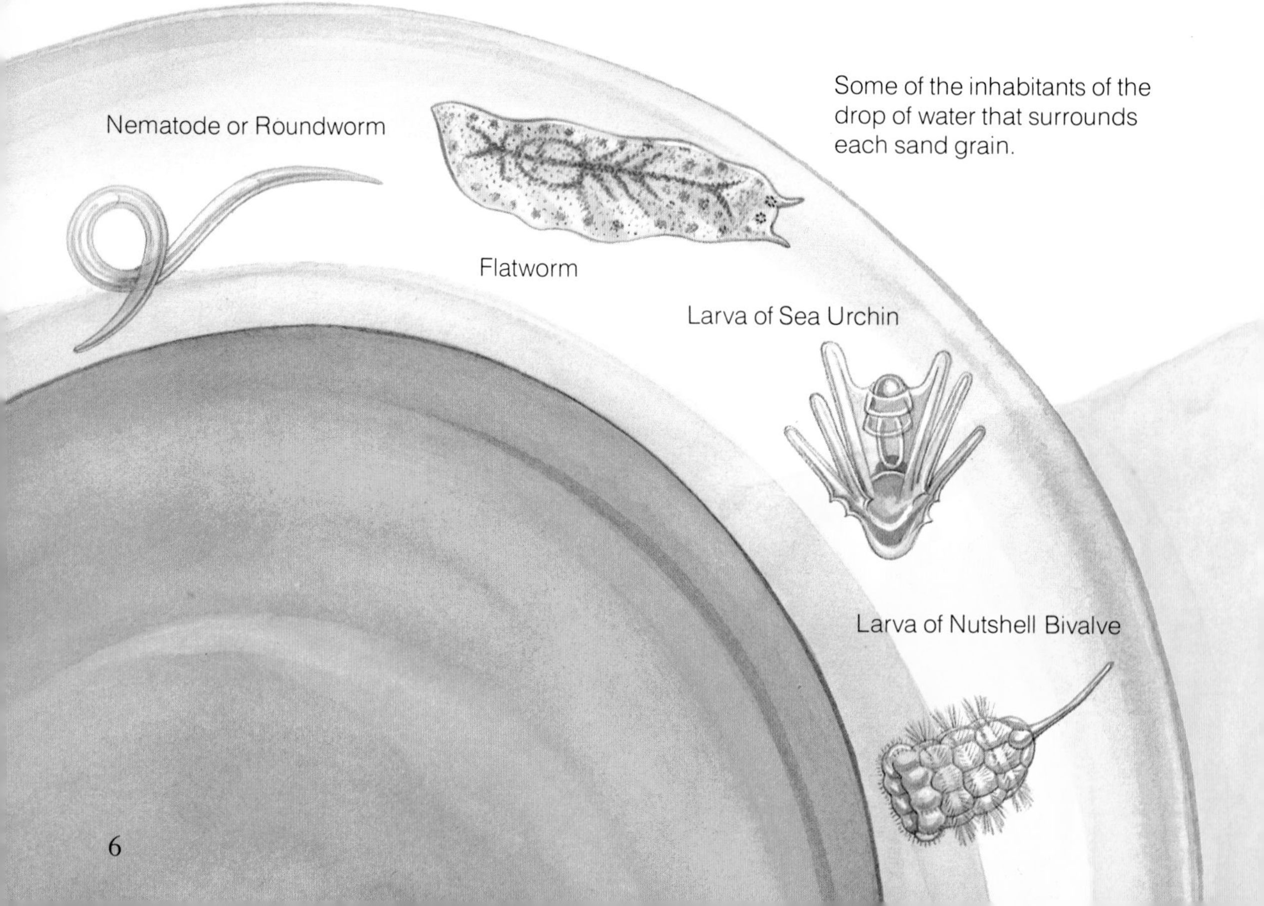

Some of the inhabitants of the drop of water that surrounds each sand grain.

Nematode or Roundworm

Flatworm

Larva of Sea Urchin

Larva of Nutshell Bivalve

Many of the animals of the shore are armored with shells. This is partly to defend them against flesh-eaters, but also to protect them against the power of the sea. When a storm is blowing, the tide may come crashing in with enough force to destroy anything which is not well-shielded. Shells also act as sunshades against the drying power of the sun and wind at low tide.

The seaweed and other plants of the shore are protected, too. Although they look fragile, they have to be strong enough to stand the force of storms that would destroy a cedar tree. At low tide, they drape over the rocks and are often slimy and dangerous to walk on. This is because the fronds are covered with a slippery gel which prevents the plant from drying out.

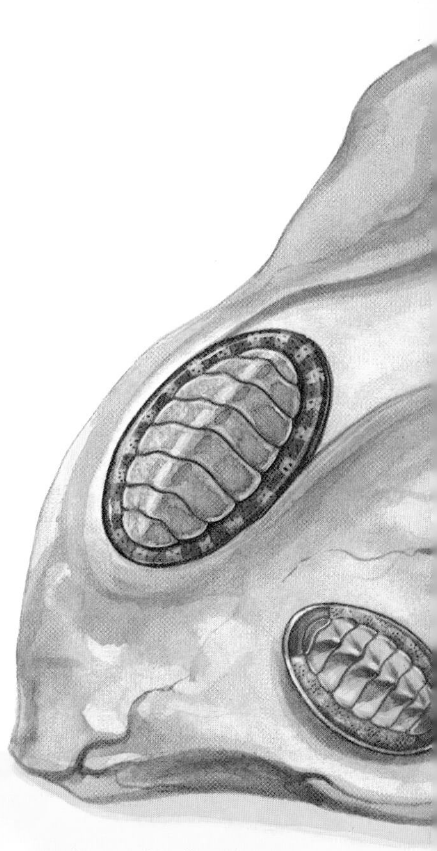

Chiton Shells

Masked Crab

Animals in Armor

The animals on the seashore wear many different kinds of armor. Sea snails have whorled shells which are heavier and thicker than those of most land snails. Many have a back door, called an *operculum*, which allows them to close the shell completely for protection.

Crabs and their relatives the prawns and shrimps have shells made of a substance rather like our fingernails. It is jointed so that they can move easily, but it does not grow as the animals get bigger. A crab has to shed its shell many times during its life. When it is ready to do so, a split appears and, in about two minutes, the crab wriggles out of its old armor. Every hair is pulled clear and even the covering of its eyes is removed. Underneath is a soft, stretchy skin which hardens in a few hours or days to form a new shell. Until this happens, the crab has to stay in hiding.

Rough periwinkles live high up on the seashore. They breathe with lungs, not gills.

Cast shell of a shore crab. This will lie on the shore until it is broken up by the waves.

Crab in hiding with a new, soft shell.

Calcareous Tube Worm

Many sea worms live in tubes which form both their home and armor. In some cases, the tubes are made of hard chalky material. In others, the worm cements sand grains together to make a strong protective tube. In spite of their armor, the worms are among the most wary of animals. The tubes may look empty – but if you are very patient, you may see a crown of tentacles shoot out from a tube to grab food and oxygen from the water.

Sand Mason Worm

Feather Duster Worm

What Do Seashore Animals Eat?

Seashore animals need to eat, just as the creatures of the land do. But the seaweed of the beach is not easy to digest and, apart from some of the sea snails, few shore-dwellers eat them. Look at oar weeds that have washed up on the beach and see if you can find the beautiful little blue-rayed limpet. This has a file-like tongue with which it rasps a notch in the stem of a kelp plant. Here it lives, protected and surrounded by food. Other animals are flesh-eaters, catching and feeding on creatures smaller than themselves. These include sea anemones and many seashore fish. Some, like crabs, are garbage-collecters, doing a useful job finding and eating the remains of dead plants and animals.

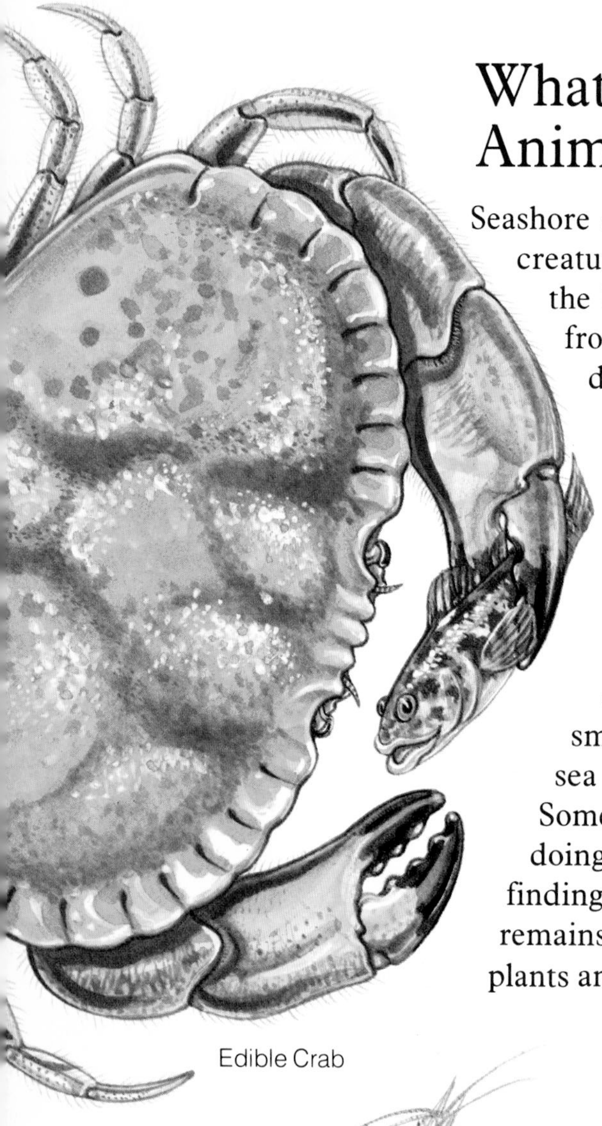

Edible Crab

Hungry sea anemone with tentacles expanded.

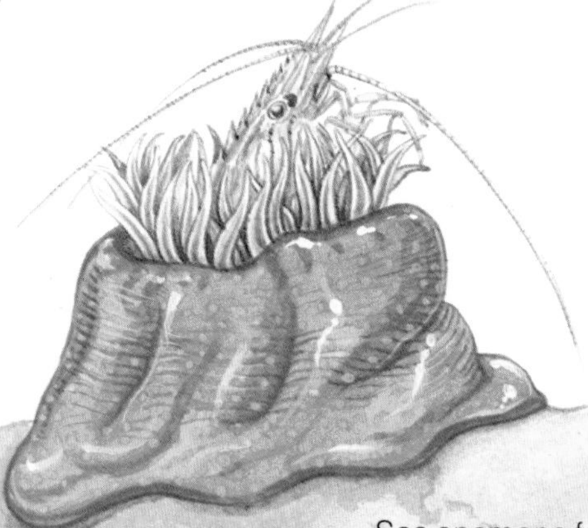

Sea anemone feeding on a shrimp.

A way of feeding not found on land is called *filter feeding*. Animals that eat in this way rely on the millions of tiny plants and animals carried in seawater. These are called *plankton*. A filter-feeder takes water into its body and strains it to catch the plankton. It is a good way of feeding, for there are huge numbers of plankton animals. Also, since the water contains oxygen, many plankton-feeders save energy by using the same movements to feed and breathe. Cockles and most other bivalves are filter feeders. So are sea squirts.

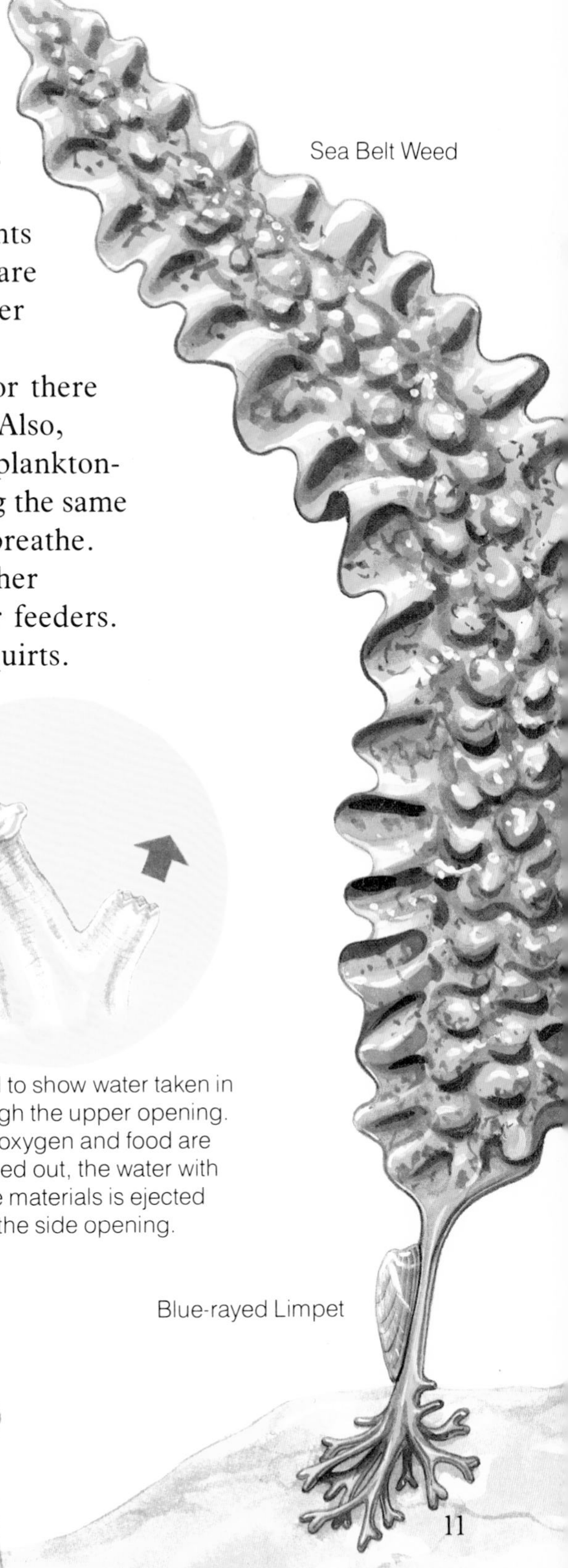

Sea Belt Weed

Detail to show water taken in through the upper opening. After oxygen and food are strained out, the water with waste materials is ejected from the side opening.

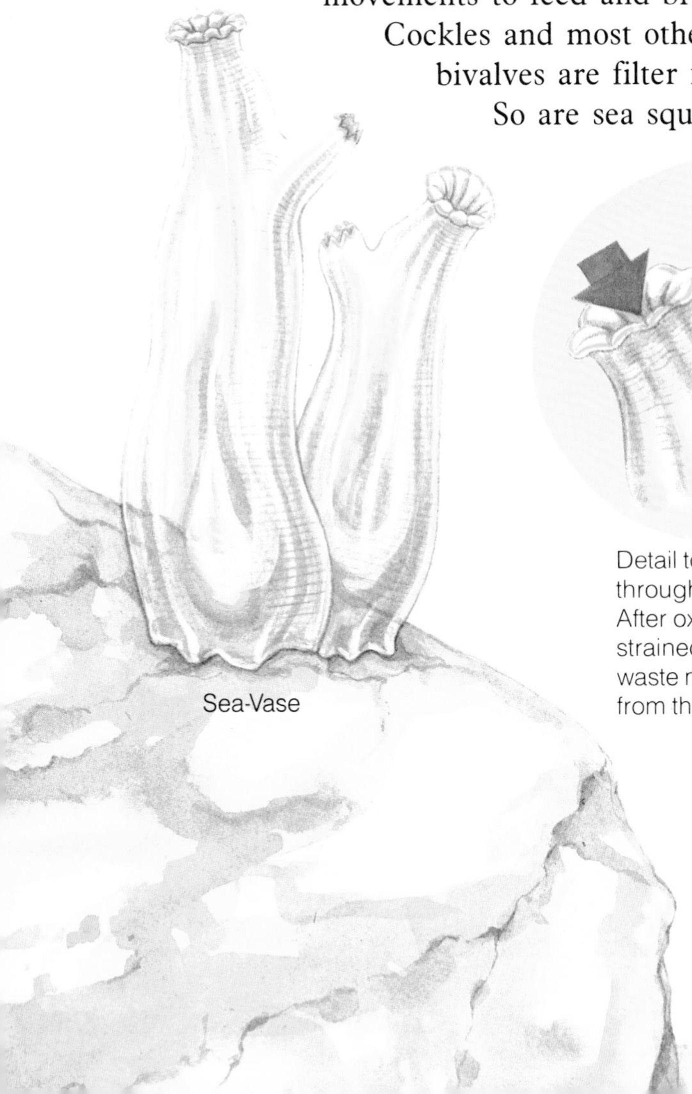

Sea-Vase

Blue-rayed Limpet

Life on the Rocks

Most people think rocky shores are the most fun to visit. You can scramble over the rocks, and there may even be caves to explore. But Nature Club members will always remember the tide. When the tide comes in, it can leave you trapped, especially on a rocky shore.

You will find the softer and more fragile animals where rocks are sheltered. Very little can survive where the shore is open to the sea and the full force of the waves. But look for limpets and barnacles in these more exposed places, for they can stand the pounding of the storm waters.

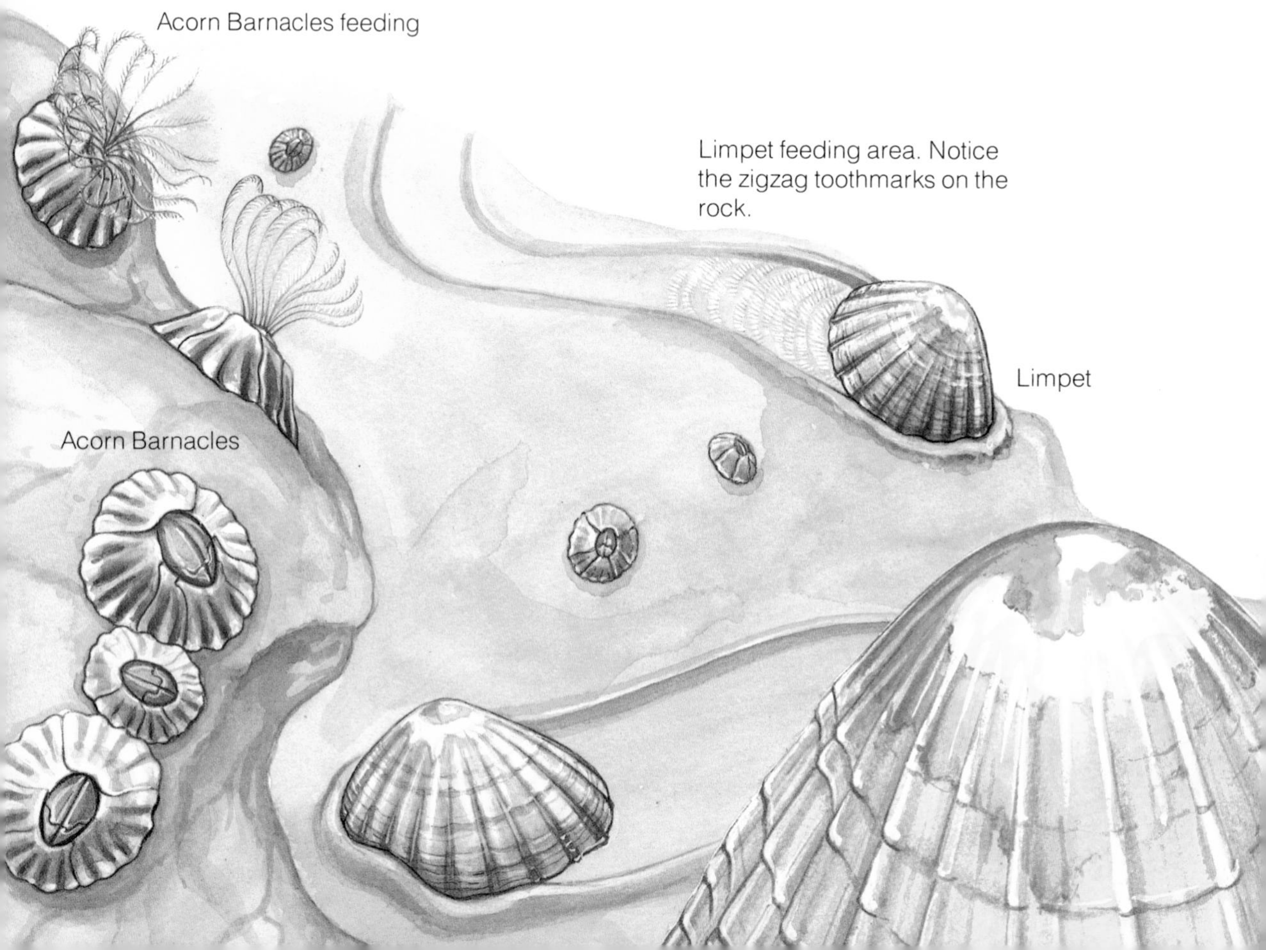

Limpets are snails, though their shells are cone-shaped rather than coiled. They live in one spot, and move up to 12 inches from it in order to feed on different kinds of small seaweed. Often the rocks are almost bare of any kind of growth because the limpets have removed it all. Sometimes the only weeds are those growing on the limpets' own shells. In some places you can follow the limpets' trail, where they have grazed away from and back to their home spot. You may even be able to find telltale zigzag marks cut into the rock by their grinding teeth.

Barnacles cling to the rocks and can't move away to look for something to eat. But when the tide washes over them, they open up the limy plates that cover their bodies and kick out huge hairy feet to comb oxygen and tiny fragments of food out of the water.

Blue-rayed limpets live in the safety of notches that they eat in the stems of seaweeds.

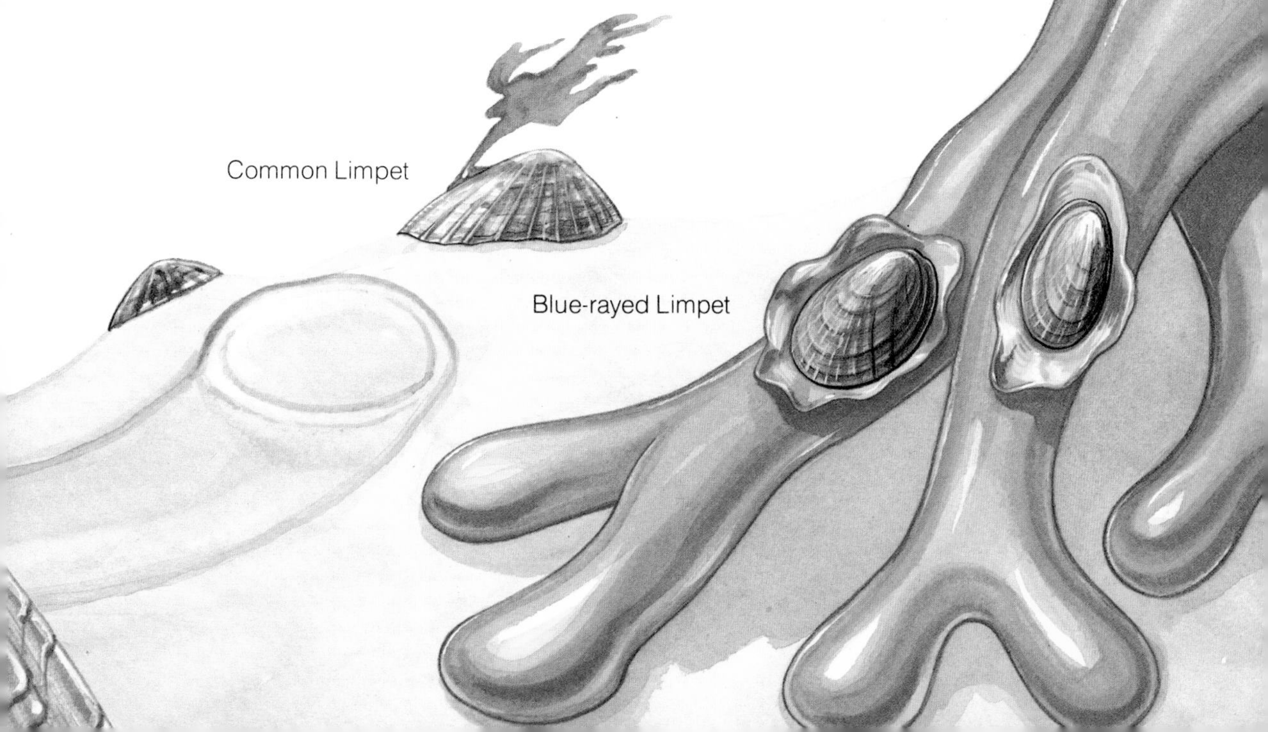

Common Limpet

Blue-rayed Limpet

Life in Tide Pools

A rocky shore is uneven, and when the tide goes down some water gets trapped in little pools that rarely dry out. Look into a tide pool and you will see animals that cannot survive if they become dry. They are usually able to stand big changes in temperature. On a hot day the water in the pool becomes very warm, but when the tide splashes back again the temperature may drop 70°F in seconds.

The walls of the tide pool are often splashed with pink color. This is not part of the rock, but a living plant spread over it. It is a kind of seaweed that is able to harden itself with calcium and other minerals, which it takes from the seawater. You would need a microscope to observe the living parts of the plant, but if you were to watch and measure it every day for a long time you would see it change.

This seaweed looks like a small green feather.

Goby

Look with your magnifying glass for the little crab-claw tips on the fronds of this weed.

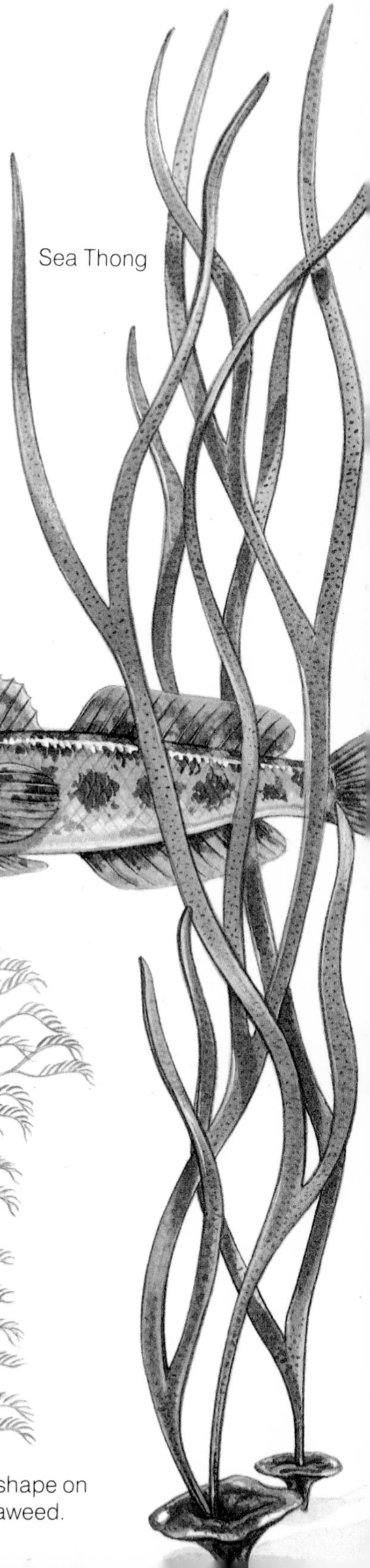

Gut Laver

You may find other seaweed growing in the shelter of the tide pool. Some may be green, or brown, but the most delicate ones are dark red in color. See how they are supported and float up in the water in the pool.

In the pool itself you may see other things that look like flowers, with many narrow petals. These are not plants, but animals called sea anemones. The "petals" are in fact tentacles armed with sting cells, for sea anemones are flesh-eaters that catch tiny fish and shrimps.

Sea Thong

Sea Lettuce

Look for the zigzag shape on the fronds of this seaweed.

Animals in Tide Pools

If you look quietly into a tide pool you can watch the behavior of all sorts of creatures. A periwinkle or whelk may plow across the floor of the pool. They do not leave footprints, but groove marks where they have been. A crab may scuttle sideways on the tips of its toes. It leaves a lacelike track, showing where it has walked.

If you dangle your toes or fingers into the warm water of the pool, shrimps and prawns may come out to investigate. They feel the shape of your hand or foot with their long antennae. If you move quickly or try to catch them, they flick their tails and shoot backward into the shelter of a crevice or some seaweed. They are very difficult to see, for their bodies are almost transparent.

Bladder Wrack

Sea Urchin

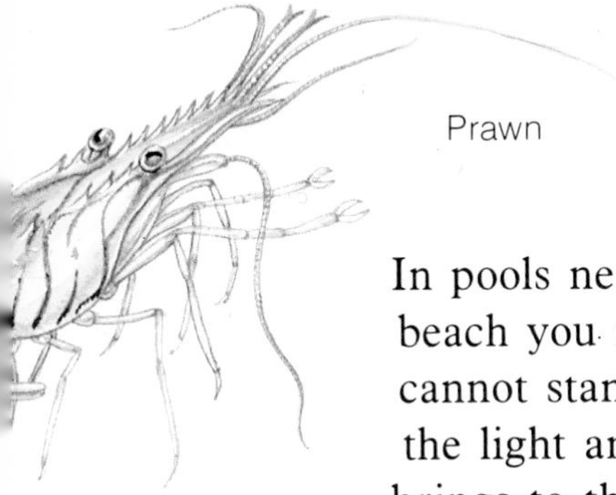

Prawn

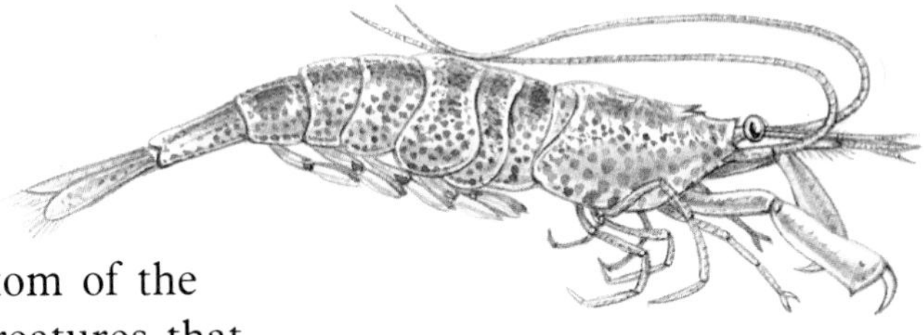

Shrimp

In pools near the bottom of the beach you may see creatures that cannot stand any drying out, or even the light and warmth that low tide brings to their secret homes. Look for sea urchins, but be careful of their spines, which are very sharp and brittle.

In bright light, sea urchins often catch a piece of floating seaweed or a flat bit of stone to hold over themselves, like a sunshade.

A shy inhabitant of pools on the lower beach is the octopus. It stays in hiding during the daytime but hunts for crabs at night. Often you can find an octopus' lair by the pile of crab shells that are the remains of its meal.

Octopus outside its lair.

Life Under Stones

A rocky beach often has boulders strewn across it. Sometimes they are huge, but there are always smaller stones, too. If you lift one of these carefully you may find all sorts of animals hiding underneath it. Crabs often burrow under stones. The velvet swimming crab is one that does this. Be careful if you pick one up, for it will defend itself with its claws and can give you a bad scratch. Sometimes you can find what looks like a sea-snail shell under a stone. If you look more closely, you will discover that the shell is the home of another sort of crab. This is called the hermit crab. Unlike its relatives, which are well protected with hard armor, the hermit has a soft tail. It protects itself by tucking its tail into an unused shell. Most hermits are not shy. Hold one for a moment, and it will emerge and scuttle across your hand.

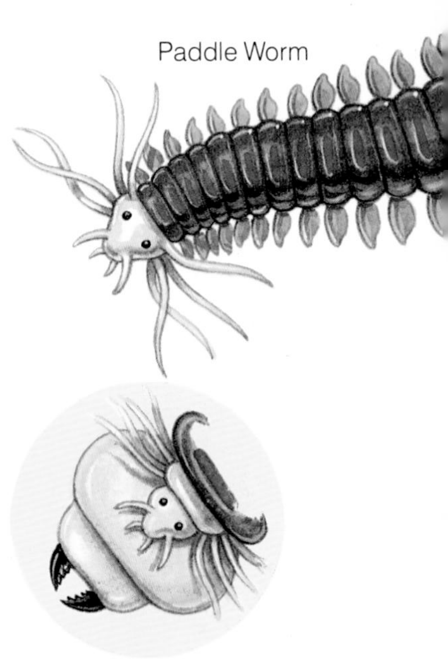

Paddle Worm

Detail of jaws of paddle worm.

Hermit crab in a whelk shell carrying a sea anemone. The anemone's stinging tentacles protect the crab from octopuses.

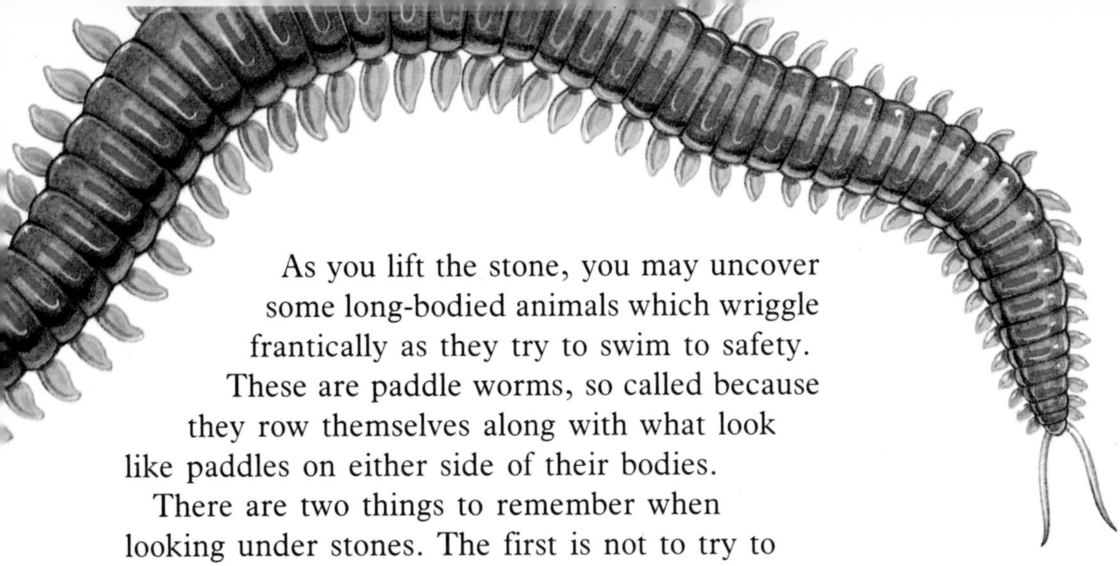

As you lift the stone, you may uncover some long-bodied animals which wriggle frantically as they try to swim to safety. These are paddle worms, so called because they row themselves along with what look like paddles on either side of their bodies.

There are two things to remember when looking under stones. The first is not to try to pick up anything too big. Interesting creatures live under little stones as well. The second is always to put the stone back gently, exactly as you found it. You have broken into the animals' homes and they will probably die if you are thoughtless.

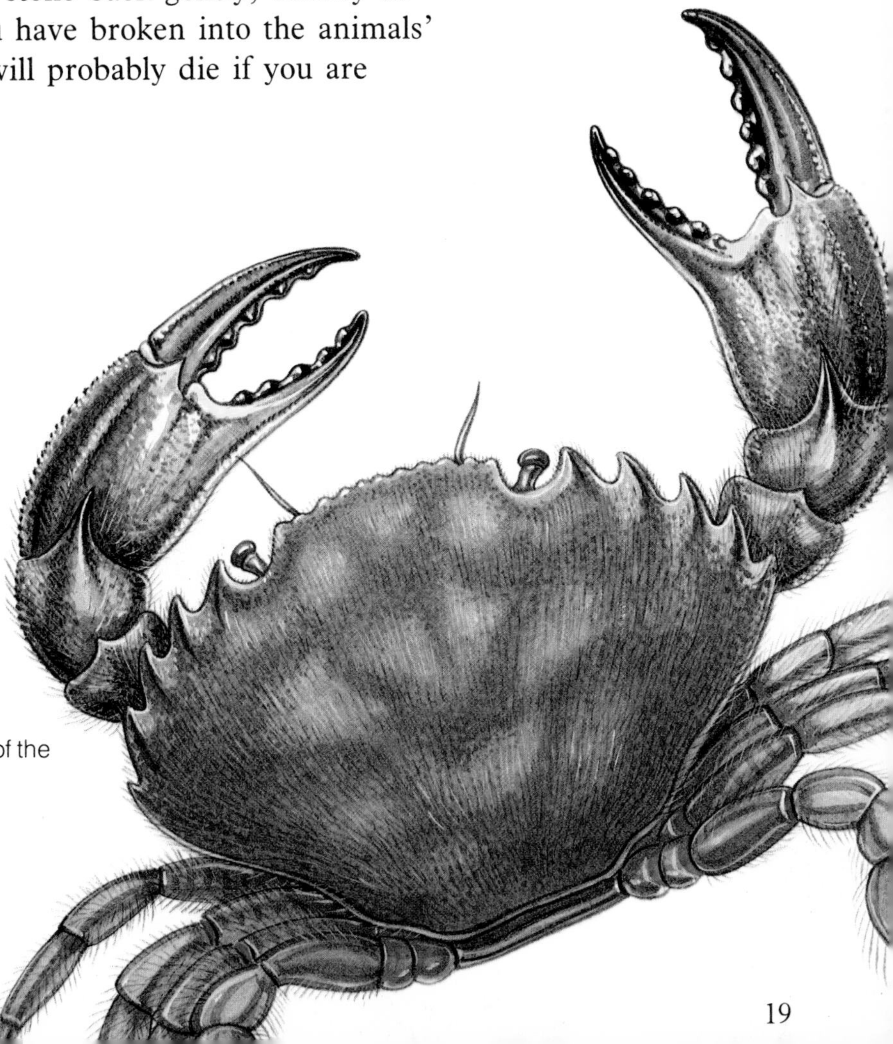

The shell of this crab protects the soft body of the animal. It is like an outside skeleton.

Encrusting Animals

Many of the tiny creatures of the seashore live in groups, or colonies. Some are called *encrusting animals*, as they coat the rocks or weeds with a living crust. If each animal lived alone it would be quite difficult to see, though usually the colony is big enough to find easily.

One common encrusting animal is the sea mat. This makes a kind of lacework pattern over stones or fronds of seaweed. Look closely and you will see that the colony is like a huge housing development, made up of hundreds of apartments. The tenant of each one is a little creature with tentacles around its mouth. It feeds on even smaller animals – plankton brought to it by the sea at high tide.

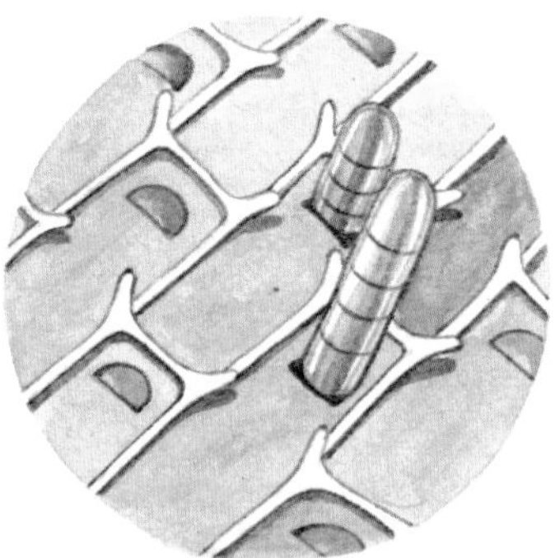

Sea mat seen through your magnifying glass showing "tower" growth.

Sea Mat

Another encrusting animal is the golden star sea squirt. This makes a jelly-like covering on seaweed or shells or rocks. In it you can see star-shaped colonies consisting of several animals. Each animal has its own mouth, so it takes its food from the seawater. But the neighbors share a main drainage system where waste products are pumped out. A very different kind of encrusting animal is the crumb-of-bread sponge. This has many pore-like mouths, but is a single animal that spreads its colored rubbery substance over the rocks.

Bread crumb sponges may be many different colors, usually pink, green or yellow. Also, their shape varies as they sprawl over the rocks.

The golden star sea squirt may also be different colors, but the "stars" always contrast with the background.

Life Among the Seaweed

Most people dislike the seaweed that drapes the cliffs and boulders on a rocky shore. But when you look at it closely, you will find that it is often a beautiful color and shape. You can see this best as it floats underwater in a tide pool.

Seaweed is home for many sorts of animals. Most of these are small, so members of the Nature Club who are exploring the world of the seaweed will need to have their magnifying glasses with them.

Lift up the curtain of seaweed growing over a rock. The first thing you see will probably be a lot of little creatures scurrying to find a safer shelter. Use your net to catch some of these so that you can see them better. Some of them are tiny crabs, too small to risk life in a tide pool or the open sea Others are little shrimplike animals.

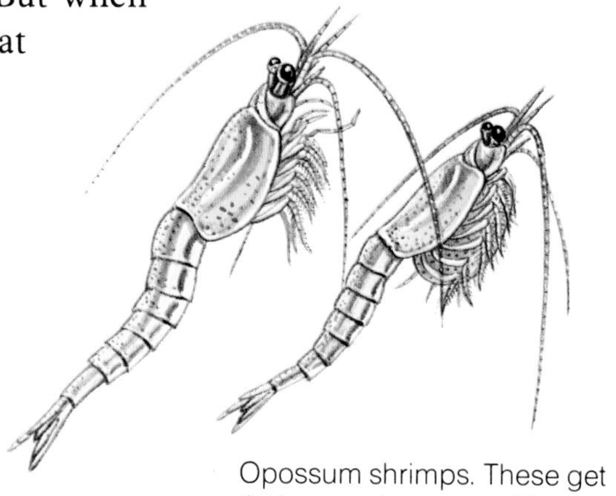

Opossum shrimps. These get their name because of the brood pouch carried by the females.

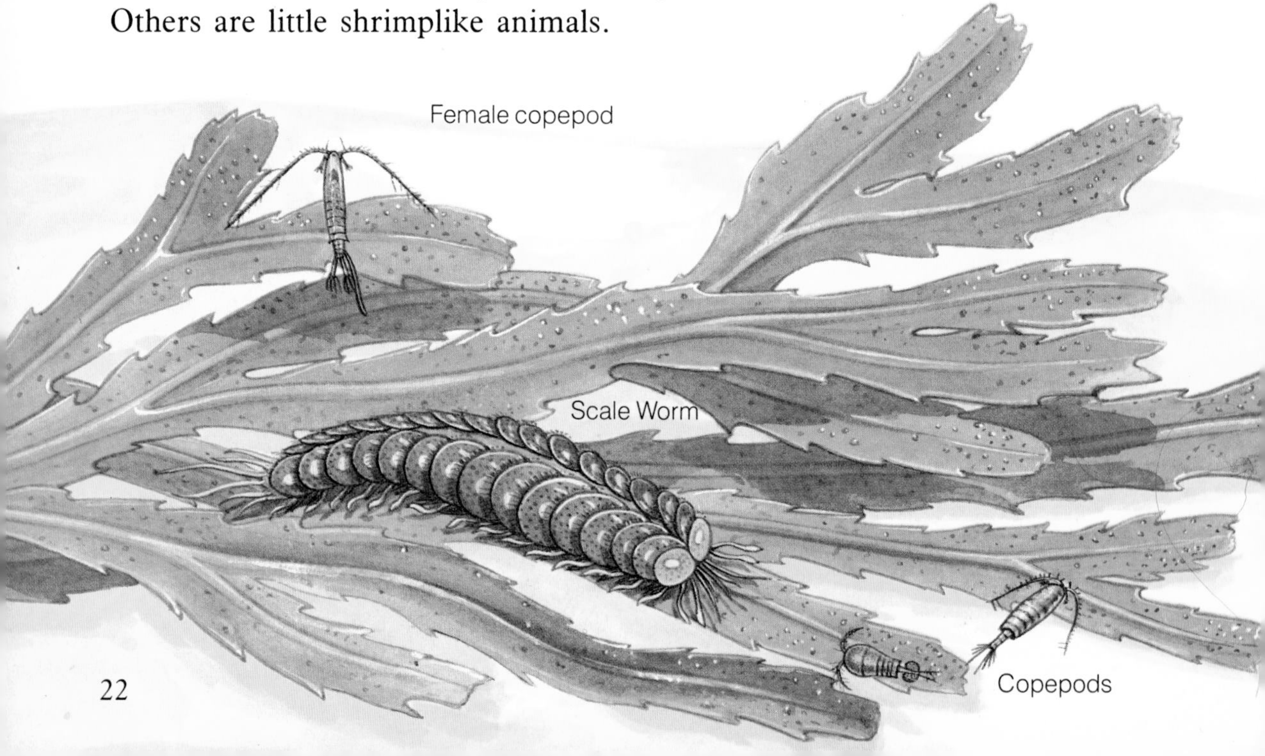

Female copepod

Scale Worm

Copepods

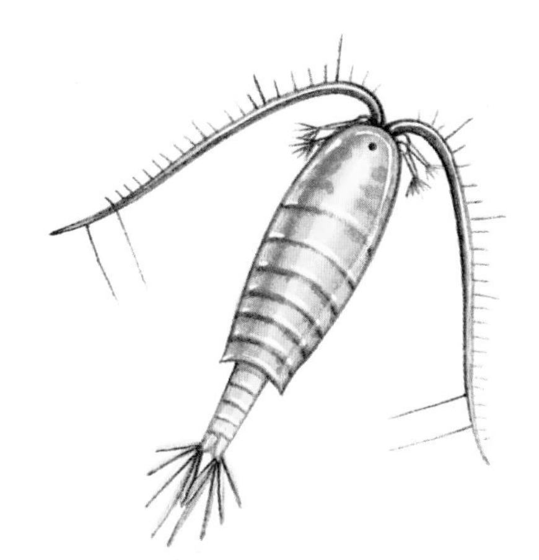

Calanid copepods are small creatures found among the seaweed.

Copepod carrying egg sac.

Skeleton Shrimp

One to search for in particular is the skeleton shrimp. This is difficult to see, because it looks very much like a small piece of seaweed. Opossum shrimps are also almost invisible against the weed, though in water their dark eyes may show up, giving a clue to their presence.

Sometimes larger animals hide under the seaweed. For instance, you may find a crab that has just changed its shell. It needs shelter until its new armor has hardened and it can face its enemies once more.

Idotea, a sea-living relative of wood lice.

Sponge

Bootlace Worm

Sea Spider

Life on Sandy Shores

At first sight, a sandy beach seems to have no life on it at all, except maybe some vacationers and a few gulls resting in the distance. Yet in the darkness beneath the sand many kinds of creatures live unseen. Some of the animals of the shore are tiny – so small that they can live in the film of water that surrounds each sand grain.

But there are many bigger animals, too. Fishermen dig in the sand because they know that fat lugworms live there. Each worm lives in a tube shaped like the letter U. At low tide you can see the coiled worm casts that lie above the tail ends of the tube, and the little pitholes above the lugworms' heads. The worms eat the sand, and digest the tiny creatures that live in it.

Other animals below the sand are *bivalve shells*. When you dig in the sand you don't usually find them, for many of them can dig down faster than you can.

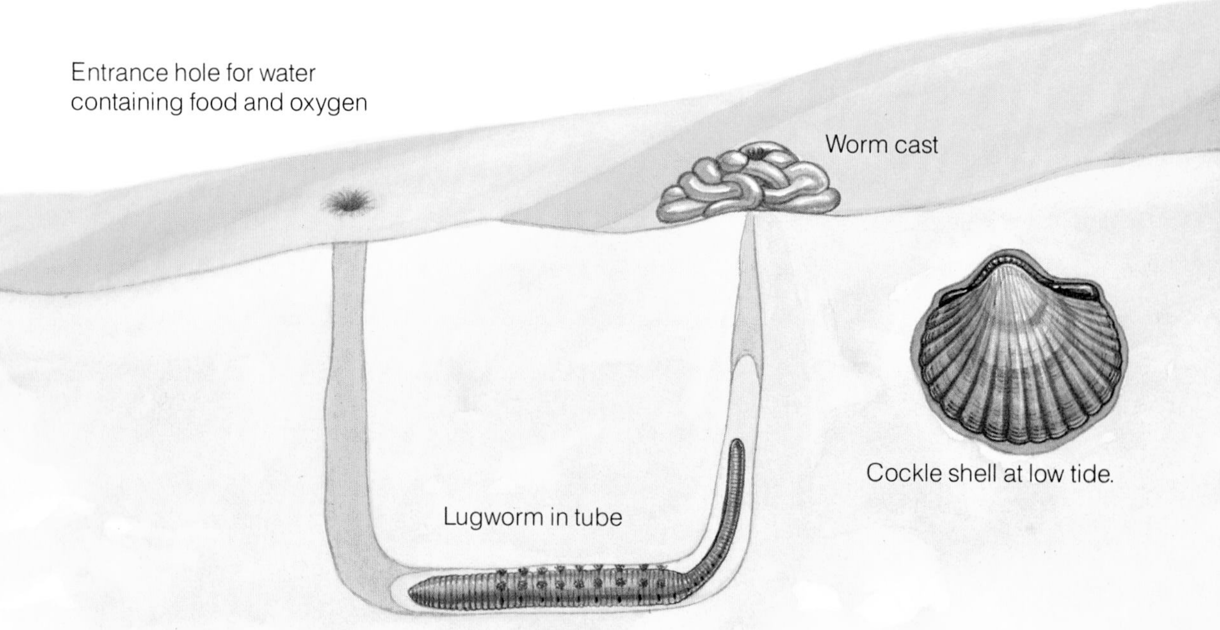

Cockle shell at low tide.

At low tide they are resting, but as the water rises the bivalves push two tubes called siphons up into it. Water containing oxygen and plankton enters one of the tubes. As it passes through their bodies, the oxygen and food are strained out. It is finally pumped out through the other siphon, along with any waste products. Bivalves with rounded shells, such as cockles, do not burrow deeply. Those that are flatter, such as wedge shells, live lower down and have much longer siphons to reach the surface. So there are layers of shells and, as a result, on many sandy beaches there may be more than 5,000 bivalves to every square yard.

Cockle feeding at high tide.

Top shell (may be found on surface of sand near rocks or seaweed).

Turrid shell

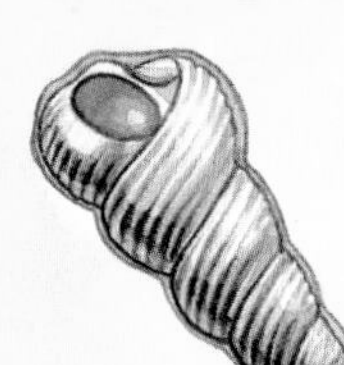

Razor Shell

Sunray Shell

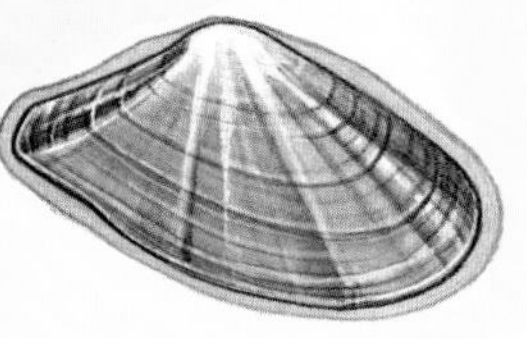

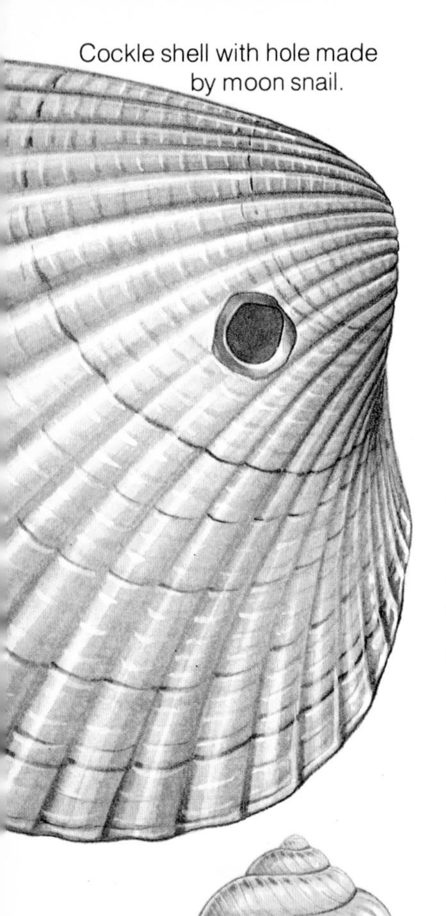
Cockle shell with hole made by moon snail.

More About Sandy Shores

Even the hidden diggers have their troubles. Some long-beaked birds, such as oystercatchers, (which don't catch oysters), can pull them out and open their shells. Another enemy which feeds on bivalves is the moon snail. This is a rounded snail that is able to move easily through the sand. When it comes across a bivalve, it uses its file-like tongue to rasp a hole through the shell of its prey, then feeds on the soft flesh inside. You can often find shells with a tiny hole neatly bored in them. This is the work of the moon snail.

Necklace Shell

The sand is not flat, and many creatures live in the pools that form across it. Shrimps lie buried just below the surface, and the young of various kinds of flat fish, such as flounders and dabs, live there, too. These young fishes are about an inch and a half across. They are very difficult to see, for they are able to change color to match their background, so they are invisible to birds and other creatures that might like to eat them.

This fish lies against the shelter of rocks.

Flounders are often in brackish water, where a river runs across the sand.

Plaice return to the sea as they grow larger.

Animals from the Open Sea

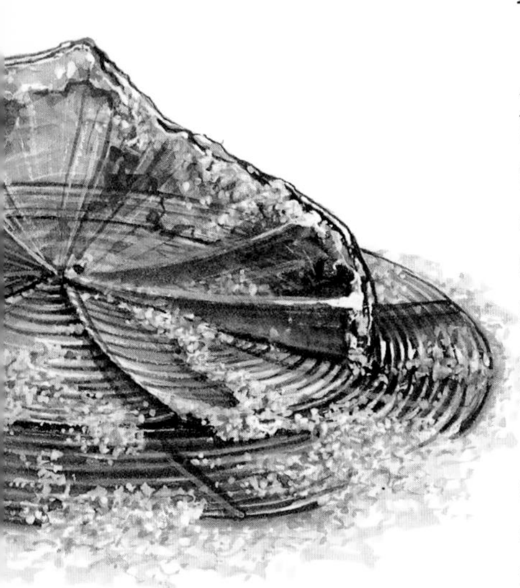

By-the-wind sailor. The horny sail of this animal, which is a member of the plankton, catches the wind. It is sometimes blown ashore.

Although we sometimes think of the shore as being part of the sea, it is really very different from the open ocean. Animals from there soon die if they are washed up on the beach. We sometimes hear of this happening when a whale gets stranded. Fortunately, whales don't often get stranded on the beach, but many smaller animals from the ocean may be found there.

One is a jellyfish. Jellyfish have no power to swim against the strength of the tide or currents, and sometimes end up on the beach. In spite of their fragile appearance, jellyfish are poisonous animals and you should not touch them, for the sting from the tentacles of most kinds is very painful. Worse is a relative of the true jellyfish, a creature called the Portuguese man-of-war. If this stings you, it will ruin your vacation. So if you hear of these animals being in the water near a beach you want to visit, the smart thing to do is to change your plans and go somewhere else.

The sea slug and its spawn are found in rock pools in the late spring. The complete spawn mass looks like a tangled ball of wool.

Portuguese Man-of-War

Some animals come into shallow water to lay their eggs. Many of the shrimplike creatures do this, and so do some of the sea slugs. Like land slugs, sea slugs are just snails without a shell. But they are often strangely shaped and brilliantly colored. They lay tens of thousands of tiny eggs, called *spawn*, that float in the water. When the eggs hatch, the young larvae look nothing like their parents. They are swept out to sea with the tide and become part of the plankton until they change and grow up.

One of the things to look out for on the shore is a mermaid's purse. This is the covering of an egg laid by a skate or small shark. After the fish hatches, the empty case, which is quite hard and strong, often gets thrown up on the beach. It is a good thing to collect to remind you of the pleasures of beachcombing or as a souvenir of a seaside vacation.

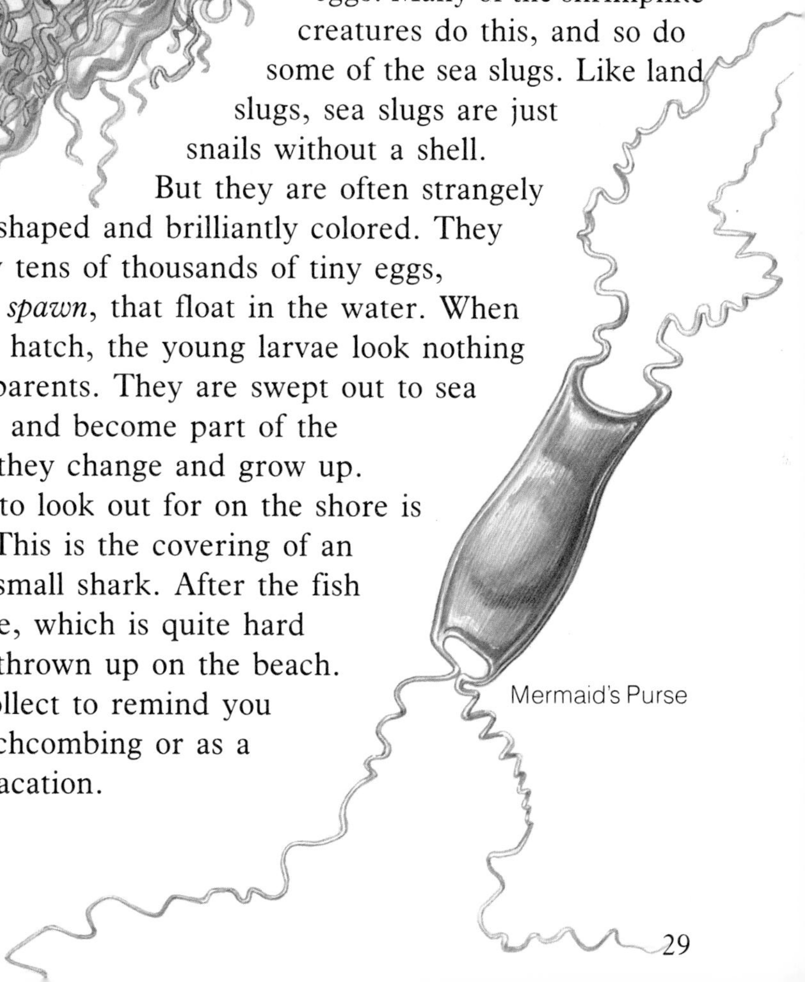

Mermaid's Purse

Glossary

bivalve shell shells made of two hinged parts. Animals that have bivalve shells feed by sucking in water containing food. Oysters and clams are typical bivalves.

colony a group of seashore animals which have developed from a single individual which has grown and developed to make many more. Usually they are connected.

encrusting animals small creatures which form a layer or crust of living animals over rocks, breakwaters or seaweed. Many kinds of sponges are encrusting animals; so are some sea mats and sea squirts.

filter feeding the method many sea creatures use to get food. Filter feeders suck in a lot of water to strain it for tiny plants and animals. Many animals without backbones make sieves of their gills, so they use the same action for feeding and breathing.

operculum a small cover some snails use to close the entrance to their shells when they are resting.

plankton the name given to plants and animals that float about in the oceans. They are carried by the water, but cannot swim against tides and currents. Most are tiny, and many are the young stages of seashore animals.

tide pool a place where a little sea water remains when the tide has gone down, often in a crack among the rocks. Seaweed and other plants may survive there, and in the shelter they provide all sorts of creatures may be found.

spawn the eggs of small animals which live in water. Animals without backbones which live on the seashore often produce huge numbers of eggs in a spawn mass.

spring tides the highest and lowest tides, which occur when there is either a new or a full moon.

Index

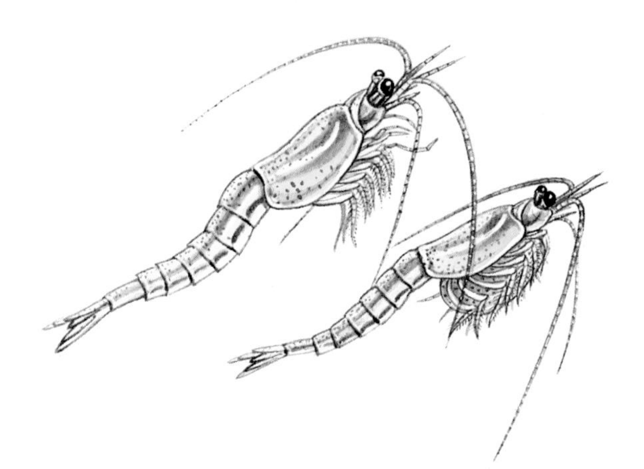